The Determination of Sulphur Dioxide in Foods Containing Volatile Oil of Mustard

THE DETERMINATION OF SULPHUR DIOXIDE IN FOODS
CONTAINING VOLATILE OIL OF MUSTARD

BY

HUBERT   MINTON   ROSENCRANS

A   THESIS   SUBMITTED   FOR   THE   DEGREE

OF

MASTER OF SCIENCE

UNIVERSITY   OF   WISCONSIN
1915

T A B L E OF C O N T E N T S

THE DETFFMINATION OF SULPHUR DIOXIDE IN FOODS
CONTAINING VOLATILE OIL OF MUSTARD

The pungent taste of the root of horse-
radish,Cochlearia amoracia L., is due to a volatile
oil, the formation of which is caused by the presence
of a glucoside sinigrin.[1] When the juicy root is
grated, the glucoside is broken up by an enzyme,
myrosin,with the formation of this oil.

Boutron and Fremy [2] (1840) found that
volatile oil of mustard is produced by the action of
an enzyme on the sinigrin found in the seeds of the
black mustard and Bussy [3] (1840) named this enzyme
"myrosin."

For a long time it was uncertain just how
the myrosin acted on the sinigrin but Gadamer [4]
(1897) finally removed the last uncertainty and
showed that the hydrolysis takes place by the addition
of the elements of one molecule of water. He also
showed that the formula for sinigrin is
$C_{10}H_{16}NS_2KO_9$ and not $C_{10}H_{18}NS_2KO_{10}$ as was supposed by
Will and Koerner. The reaction can be expressed as

follows:

$$C_{10}H_{16}NS_2KO_9 + H_2O \text{ (myrosin)} = S \ C \ N \ C_3H_5 + C_6H_{12}O_6 + K \ H \ S \ O_4.$$

    Sinigrin   Allyl mustard oil.

It was first shown by Hubataka [5] (1843) that the
volatile oil produced in the root of horse radish
is identical with that produced in the seed of black
mustard and this was later verified by G. Sani [6] in
1892 and by Gadamer [7] in 1897. Both of these men showed that the oil was produced by the action of the
ferment myrosin on the sinigrin found in the root,
the formation of the oil being caused in the same
manner as in the seed of the black mustard.

        This volatile oil can also be obtained from
other cruciferae.It can be obtained from the leaves
and seeds of shephard's purse(Capsella bursa pastoris),
hedge mustard(Sisymbrium officinale),penny-cress
(Thlaspi arvense), iberis(Iberis amara), and in the
roots of some species of acacia.Other mustard oils
(isothiocyanates)also occur in nature. Oil of garden-
cress(Lepidium sativum L.)has been identified as
benzyl isothiocyanate, $C_6H_5CH_2NCS$, while a volatile
oil obtained from scurvy grass(Amoraciae radix) has

been shown to contain secondary butyl isothiocyanate,

$$\begin{matrix} C_2H_5 \\ CH_3 \end{matrix} \!\!\!\! > \!\! CHNCS .$$

The composition of volatile mustard oil was first studied by Will[8] (1844) and simultaneously by Wertheim [9] (1844). They regarded it as allyl thiocyanate and gave it the formula $C_6H_5C_2NS_3$. Later, however, Will with Koerner[10](1863) suggested the formula C_3H_5NCS which was confirmed by Billeter and Gerlich[13] (1875). These investigators recognized the true constitution of volatile mustard oil as the allyl ester of the isothiocyanic acid and showed that by the interaction of allyl iodide and potassium sulphocyanate, allyl sulphocyanate is first formed and that upon heating, this is converted into its isomer, the allyl isosulphocyanate.

Hofmann [11] in 1868 had already suggested as an explanation of this form of isomerism that in the true thiocyanates, the carbon of the allyl radical is directly united with the sulphur, but in the "iso" compounds with nitrogen.

The possibility that allyl sulphocyanate

might first be formed by the ferment action of
sinigrin was excluded by Schmidt [15] when he allowed
this action to take place at a low temperature. He
found that even at $0°$allyl isosulphocyanate is
formed and only traces of the normal isomer.

      Dumas and Pelouze [13] (1833) showed that by
the action of ammonia on volatile mustard oil, a
crystalline product is obtained. This was verified
by Will [14](1844) who gave to the new compound the
name"Thiosiamine" and made use of its formation in
the first quantitative determination of the oil. He
converted the oil into the thiosiamine and then by
treating this with litharge,obtained all of the
sulphur as lead sulphide and from the amount of this
calculated the quantity of mustard oil.

      Many methods have since been worked out
for the quantitative determination of volatile mustard oil. Most of the early investigators precipitated
the sulphur as a metallic sulphide and calculated the
mustard oil from the weight of this precipitate.

      Kremel [16] (1888) suggested the use of
ammonia of definite strength and determined the excess
with half normal acid. However, no practical use of this

method seems to have been made.

Jorgensen [17] (1898) distilled the volatile mustard oil into 20 cc. concentrated ammonia and 50cc. alcohol. This mixture was placed on the water bath over night, the ammonia expelled and then the nitrogen determined by the Kjeldahl method. Two atoms of N are equivalent to one molecule of mustard oil.

According to Gadamer [18] (1899) the mustard oil can be determined by converting it first into the thiosiamine, then adding N/10 silver nitrate and titrating the excess with ammonium sulphocyanate, using ferric chloride as an indicator. Grutzner [19] converted the allylisothiocyanate into thiosiamine and oxidized this with sodium peroxide. The resulting sodium sulphate is determined as barium sulphate.

Dieterich [20] (1886) converted the volatile mustard oil into the thiosiamine and then determined the sulphur as silver sulphide. He added silver nitrate to the thiosiamine, washed, dried and weighed the Ag_2S, from which he calculated the mustard oil. Foerster [21] got better and quicker results by using freshly prepared mercuric oxide and weighing the

mercuric sulphide which was precipitated.

Schlicht [32] proved, however, that by this method of Foerster there was always a loss of 3.4 to 7.2 %. He also tried the method worked out by Dircks [23] in which the volatile oil is oxidized by potassium permangante in alkaline solution, the excess of permanganate removed by the addition of hydrochloric acid and the oxidized sulphur of the oil determined as $BaSO_4$. He found that somewhat low results were obtained by this method and that the method was otherwise objectionable on account of the abundant evolution of chlorine and the many troublesome manipulations.

Schlicht then worked out a method [34] of his own which seems to be very satisfactory and accurate. The volatile mustard oil is distilled into alkaline permanganate and the mixture vigorously shaken, heat being applied at the same time. The sulphur is oxidized to sulphuric acid and potassium sulphate. After cooling the solution, alcohol is added and after standing for some time the manganese separates out as a brown precipitate which according to the author has the composition $KH_3Mn_4O_{10}$. Enough water is now added to make the volume up to 1000 cc., the solution filtered

and the sulphuric acid determined in aliquot portions.
of the filtrate as barium sulphate. In the reduction
of the permanganate by the alcohol, a part of the
potassium sulphate may also be reduced to sulphite.
To correct this a solution of iodine in potassium
iodide is added to the filtrate until a slight yellow
color remains. The sulphur is then determined as
$BaSO_4$ and the amount of mustard oil calculated from
this.

SULPHUR DIOXIDE IN FOODS

Sulphurous acid and sulphites have often been used in foods either as preservatives or as bleaching agents or both,but due to the strict food laws which we now have, it is not so generally employed. Sulphites were once used quite extensively in chopped meats and their occurrence in such products is not uncommon today.

Calcium bisulphite is the salt commonly employed and the amount of sulphur dioxide found varies from 0.01 to 0.34 %. It serves in chopped meats only to a slight extent as a preservative,acting chiefly as a deodorizer and a restorer of the bright red color of fresh meat.

Free sulphurous acid in the form of sulphur fumes is used to disinfect wine casks,to bleach and preserve dried fruits and in the bleaching of molasses. In the case of dried fruits it probably in large part, forms compounds with the sugar,while it is stated that in the wine casks it combines with the acetaldehyde of the wine forming aldehyde sulphurous acid,which is said to be relatively harmless.

The bisulphites of sodium and calcium, $NaHSO_3$ and $Ca(HSO_3)_2$, are the sulphurous acid salts most commonly employed as preservatives of fruit juices, ketchups, wines, malt liquors and meat products.

For the detection and determination of sulphurous acid in foods, there is but one good method, the distillation method, which is carried out as follows: 50 to 300 grams of the material mixed with water and acidified with phosphoric acid are distilled in a current of carbon dioxide into water containing a few drops of bromine until 150 to 300 cc. have distilled over. The excess of bromine is boiled off, and 1 cc. concentrated HCl and an excess of barium chloride added. The barium sulphate is filtered and weighed and the sulphur dioxide calculated from this.

If sulphides are present in the original sample, as is true of decomposed meat products and possibly other foods, some hydrogen sulphide will come over with the steam from the distilling flask. This hydrogen sulphide can be readily removed by passing the steam from the distilling flask through a flask containing 40cc. of a 2 $\%$ neutral solution

of cadmium chloride before allowing it to enter the
condenser. Winton advises the use of a 1% solution
of copper sulphate.

INTERFERENCE OF OTHER VOLATILE SULPHUR COMPOUNDS.

There are other volatile sulphur compounds,
sometimes found in foods,which would interfere with
the determination of sulphur dioxide by the above
method,among them the mustard oils and the thioethers.
The occurrence of the former has already been referred
to;several of the latter occur in garlic and related
plants.

When these volatile sulphur compounds are
present in food products they will distil over with
the sulphur dioxide and in whole or in part be oxidized
by the bromine to sulphuric acid and then precipitated
as barium sulphate upon addition of barium chloride.
Such compounds must therefore be removed before a
determination of sulphur dioxide can be made.

A careful search of chemical literature
having failed to show any reference to any attempts
of determining sulphur dioxide in the presence of these
volatile sulphur compounds, the study of methods for
accomplishing this result was taken up as the subject
matter of this thesis,confining myself,however, to
allyl mustard oil from horse-radish.

Two methods suggested themselves. The first, the removal of the allyl isothiocyante with cadmium chloride interposed between the distilling flask and the condenser;the second, the conversion of this compound into thiosiamine with ammonia and the subsequent decomposition of the thiosiamine into silver sulphide with silver nitrate.

EXPERIMENTAL

In investigating the first method, a 10 % solution of cadmium chloride was placed between the flask containing the grated horse-radish and the condenser, in such a way that the steam was forced to bubble through the solution which was kept at the boiling point. About a dozen determinations were made with pure freshly-grated horse-radish root which proved conclusively that while some of the volatile sulphur compounds were removed as cadmium sulphide, most of them came over unaltered so that the method had to be abandoned.

The second method gave more promising initial results and after many attempts, under varying conditions, it was found possible to so maintain the conditions that a complete removal of all organic sulphur was accomplished and all the sulphur dioxide could be converted into barium sulphate. The method as finally worked out, as applied to horse-radish is as follows:

30-40 grams of grated fresh or prepared horse-radish are mixed with 500 cc. water, made acid with 20-30cc. of 10 % phosphoric acid and distilled,

passing a slow current of carbon dioxide through the
distilling apparatus throughout the distillation.
The volatile mustard oil together with any sulphurous acid distils over and 300-350 cc. of the distillate are collected in a mixture of 50 cc.alcohol(95 %),
and 5cc. dilute ammonia (15 cc. conc. ammonia in
100 cc. water). Enough ammonia must always be present
to insure a complete conversion of the mustard oil
into the thiosiamine. In order to insure this, the
distillate should smell strongly of ammonia before
the silver nitrate is added. The 5 cc. of dilute
ammonia which I have specified is enough unless a
large amount of horse radish istaken or one which
contains an exceptionally high percentage of the
mustard oil.   10 cc, of a 10 g. % silver nitrate
solution are then added. The beaker containing the
mixture is placed in a water bath and the temperature
slowly brought up to a point where the alcohol begins
to boil off. This temperature is kept constant until
all the alcohol and ammonia have disappeared which will
require about 30-45 minutes. The beaker is then removed
from the water bath and kept at incipient boiling for
a half hour over a free flame. The silver nitrate

precipitate is filtered off and the silver in the
filtrate, present as silver sulphate and silver
nitrate,removed by precipitation with hydrochloric
acid and filtering. The sulphuric acid in the filtrate
is precipitated with barium chloride in the usual way
and the sulphur calculated as sulphur dioxide from
the barium sulphate formed.

In the above method the mustard oil is
first converted into thiosiamine by action of
ammonia and then into silver sulphide through
reaction with ammoniacal silver nitrate solution in
accordance with the following equations:

$$S \; C \; N \; C_3 \; H_5 + NH_3 = C \underset{\diagdown S}{\overset{\diagup NHC_3H_5}{\Longleftarrow} NH_2}$$

Mustard oil   ammonia   Thiosiamine

$$C \underset{\diagdown S}{\overset{\diagup NHC_3H_5}{\Longleftarrow} NH_2} + 2 \; Ag \; NO_3 + 2 \; NH_3 = C \overset{\diagup NHC_3H_5}{\underset{\searrow}{\equiv} N} + Ag_2S + 2NH_4NO_3$$

The sulphur dioxide is converted,through
ammonium sulphite into silver sulphite which upon
heating, in the presence of excess of silver nitrate
and ammonia,is changed to silver sulphide.

Equations:-

(1) $SO_2 + 2 \; NH_3 + H_2O = (NH_4)_2SO_3$

(2) $(NH_4)_2 SO_3 + 2 Ag NO_3 = 2 NH_4 NO_3 + Ag_2SO_3$

(3) $2 Ag_2SO_3 + H_2O = Ag_2SO_4 + 2 Ag + SO_2 + H_2O$

(4) $Ag_2SO_3 + H_2O = 2 Ag + H_2SO_4$

The last two equations are possible here.
The first (No. 3) is declared by some (Abegg., Hand-
buch der Anorg. Chemie), to be the one representing
the change which Ag_2SO_3 undergoes when heated in
aqueous solution. The second, (number 4) according
to other authorities, (Prescott and Johnson, Qualitative
Chemical Analysis) is the correct reaction. From my own
results I am led to believe that the first equation
represents the reaction, at least, under the conditions
I maintained; for if the solution containing the silver
nitrate precipitate is rapidly heated to boiling, there
is a loss of sulphur from the $Ag_2 SO_3$ which would not
occur if the second equation were the correct one.
This loss is shown in Table # 1.

Table Showing the Loss of Sulphur as SO_3,
Due to Too Rapid Heating of the Solution Containing
the Ag_2SO_3.

TABLE I.

Sample Number.	Amount of SO_2 Added as Na_2SO_3	Amount of SO_2 Recovered as $BaSO_4$
1	0.1932 grams	0.0889 grams
2	0.1035 "	0.0511 "
3	0.1039 "	0.0544 "
4	0.0190 "	0.0144 "
5	0.0190 "	0.0096 "
6	0.0190 "	0.0129 "
7	0.0209 "	0.0120 "
8	0.0198 "	0.0107 "
9	0.0197 "	0.0131 "
10	0.0179 "	0.0097 "
11	0.0194 "	0.0122 "
12	0.0165 "	0.0115 "
13	0.0225 "	0.0197 "
14	0.0365 "	0.0221 "
15	0.0306 "	0.0247 "
16	0.0287 "	0.0201 "

It was noticed that this loss decreased as the time of bringing the solution to the boiling point increased. So after numerous trials I found that by the method of heating which I have previously described, viz; that of slowly raising the temperature to the boiling point, the silver sulphite could be changed to silver sulphate quantitatively.

These results are given in Table # 2.

Table Showing Results of Sulphite Determinations in Horse-radish Using the Method Finally Adopted.

TABLE II.

Sample Number.	Amount of SO_3 Added as Na_2SO_3	Amount of SO_2 Recovered as $BaSO_4$
17	0.0331 grams	0.0300 grams
18	0.0220 "	0.0208 "
19	0.0085 "	0.0082 "
20	0.0278 "	0.0285 "
21	0.0150 "	0.0153 "
22	0.0151 "	0.0153 "
23	0.0109 "	0.0111 "
24	0.0189 "	0.0183 "
25	0.0278 "	0.0228 "
26	0.0135 "	0.0117 "
27	0.0035 "	0.0035 "

Blank determinations using known amounts of sodium sulphite in the distilling flask proved that under the above conditions all of the sulphur dioxide could be converted into silver sulphate.

Blank determinations were also made on pure horse-radish alone and when the barium chloride was added, after going through the complete operation, no precipitate formed. This showed that all of the sulphur from the mustard oil was removed as silver sulphide and no longer interfered with the determination of the sulphur from the sodium sulphite.

The solution of sodium sulphite employed was standardized in two ways.

First, by distilling 10 cc. in 400 cc. of water, made acid by an excess of phosphoric acid and collecting the distillate in bromine water. The bromine was boiled off and the sulphur determined as barium sulphate..

Second, by titrating with iodine and sodium thiosulphate. It was found that, in order to get accurate determinations by this method, an excess of iodine must be taken and the sodium sulphite quickly run into it. The excess of iodine is then titrated

with standard sodium thiosulphate. Each cc. of N/10
iodine used is equivalent to 0.0032 grams sulphur
dioxide. The sodium sulphite solution was found to be
very unstable and it was necessary to standardize
it each day before using it.

Different brands of "Prepared Horse Radish"
were used and all worked equally as well as the natural
root. The"Prepared Horse Radish",however,contains a
much smaller percentage of mustard oil then the
natural root. This isnot unusual for it has long been
known that the oil disappears when the root is kept
for any length of time, especially when in contact with
water.

21

S U M M A R Y

1.   The pungent taste of horse-radish root is due to
the presence of allyl mustard oil, produced by the
action of an enzyme myrosin on a glucoside sinigrin.
This mustard oil formed in the root of the horse-radish
is identical with that produced from the seed of
black mustard.

2.   A thiosiamine is formed by the action of
ammonia on mustard oil. This thiosiamine forms
silver sulphide by interaction with silver nitrate
in ammoniacal solution, which reaction can be utilized
for the quantitative determination of mustard oil.

3.   Mustard oil interferes with the determination of
sulphur dioxide in food products by the present methods.

4.   A practical method has been worked out in detail
by which even small quantities of sulphur dioxide can
be accurately determined in the presence of volatile
mustard oil.

22

R E F E R E N C E S

1. Archiv. d. Pharm., , 235, p 577.

2. Journ. de Pharm.,II 26, 39.

   Liebig's Annalen, 34, 233.

3. Journ. de Pharm.,II 26, 48, 112.

4. Archiv. d. Pharm., 235, 44.

5. Liebig's Annalen, 47, 153.

6. Accad. Linc. 1892.

7. Archiv. d. Pharm. 235, 577.

8. Liebig's Annalen, 52, 1.

9. Liebig's Annalen, 52, 54.

10. Liebig's Annalen, 125, 257.

11. Berl. Berichte, 1, (1868) 38.

12. Berl. Berichte, 8, (1875) 464, 650,824.

13. Aun. de Chim. et Phys.,II 53 (1833), 181.

14. Liebig's Annalen, 52, 1.

15. Berichte, 10, 187.

16. Pharm. Post, 21, 828.

17. Chem. Zentralblatt,II 927.

18. Archiv. d. Pharm., 237, 110, 372.

19. Archiv. d. Pharm., 237, 125.

20. Helfenberger Annalen, (1886), 59.

21. Zeit. f. Anal. Chem. 30, 647.

22.   " " "   "   30, 661.

23.   " " "   "   32, 461.

24.   " " "   "   30, 661.

Lightning Source UK Ltd.
Milton Keynes UK
UKHW020622071122
411784UK00006B/436